EMMANUEL JOSEPH

The Art of Connection, Psychology, Nature, and the Creative Spirit

Contents

1   Chapter 1: The Importance of Connection   1

2   Chapter 2: The Psychology of Connection   3

3   Chapter 3: Nature as a Source of Connection   5

4   Chapter 4: Creative Expression as a Means of Connection   7

5   Chapter 5: Connection Through Shared Experiences   9

6   Chapter 6: The Role of Empathy in Connection   11

7   Chapter 7: Connection and Personal Growth   13

8   Chapter 8: Overcoming Barriers to Connection   15

9   Chapter 9: Connection and Community Building   17

10   Chapter 10: Connection and Well-Being   19

11   Chapter 11: Connection in Times of Crisis   21

12   Chapter 12: The Future of Connection   23

13   Chapter 13: Connection Through Storytelling   25

14   Chapter 14: Connection and Creativity in the Workplace   27

15   Chapter 15: The Legacy of Connection   29

1

Chapter 1: The Importance of Connection

Humans are inherently social creatures, wired to connect with others from the moment we are born. The significance of connection cannot be overstated, as it influences our mental, emotional, and even physical well-being. Research has shown that strong social bonds can reduce stress, increase longevity, and improve overall health. This chapter will delve into the psychological and biological underpinnings of human connection and why it is essential for our survival and happiness.

The concept of connection extends beyond human relationships; it encompasses our relationship with nature and the world around us. Throughout history, humans have sought solace and inspiration in the natural world, finding a sense of belonging and purpose in the beauty of the environment. From ancient rituals to modern-day practices, we will explore how nature has played a pivotal role in fostering connection and well-being.

One of the most compelling stories of connection comes from the life of Jane Goodall, the renowned primatologist. Her groundbreaking work with chimpanzees in Gombe, Tanzania, highlighted the deep bonds that can form between humans and animals. Goodall's ability to connect with the chimpanzees on a profound level not only advanced scientific understanding but also demonstrated the universal language of empathy and compassion.

In contemporary society, the rapid advancement of technology has transformed the way we connect with one another. Social media, instant messaging, and video calls have made it easier than ever to stay in touch with loved ones, regardless of geographical distance. However, this digital age also presents challenges, as the quality of our connections can sometimes suffer in the face of superficial interactions. We will examine the impact of technology on human connection and the importance of fostering meaningful relationships in the digital era.

Connection is not limited to our interactions with others; it also involves our relationship with ourselves. Self-awareness, self-compassion, and self-acceptance are crucial components of building a strong connection with oneself. This chapter will explore the journey of self-discovery and the ways in which cultivating a healthy relationship with oneself can enhance overall well-being and connection with others.

2

Chapter 2: The Psychology of Connection

The field of psychology offers valuable insights into the mechanisms behind human connection. Attachment theory, developed by John Bowlby and Mary Ainsworth, provides a framework for understanding how early relationships with caregivers shape our ability to form connections throughout life. This chapter will delve into the various attachment styles and how they influence our interactions and relationships.

Social psychology also sheds light on the dynamics of connection, exploring concepts such as social identity, group dynamics, and the influence of social norms. We will examine how these factors impact our sense of belonging and the ways in which we navigate social relationships. Understanding these psychological principles can help us foster healthier and more fulfilling connections with others.

One fascinating story that illustrates the power of connection is the tale of Viktor Frankl, an Austrian neurologist and psychiatrist who survived the Holocaust. In his book "Man's Search for Meaning," Frankl recounts how the bonds formed with fellow prisoners provided him with the strength and resilience to endure unimaginable suffering. His experiences underscore the profound impact that connection and shared humanity can have on our ability to survive and thrive.

The concept of emotional intelligence, popularized by psychologist Daniel Goleman, emphasizes the importance of recognizing, understanding, and managing our own emotions as well as those of others. This chapter will explore the role of emotional intelligence in fostering connection, highlighting strategies for improving our emotional awareness and interpersonal skills.

Connection is not solely a human endeavor; animals also exhibit complex social behaviors and bonds. From the intricate communication systems of dolphins to the cooperative hunting strategies of wolves, the animal kingdom offers fascinating examples of connection and social cohesion. We will explore these stories to gain a deeper understanding of the universal nature of connection.

3

Chapter 3: Nature as a Source of Connection

Nature has always been a powerful source of inspiration and connection for humans. The natural world offers a sense of peace, beauty, and wonder that can rejuvenate our spirits and strengthen our bonds with others. This chapter will explore the various ways in which nature fosters connection, from solitary walks in the woods to communal activities like gardening and hiking.

One remarkable story of connection with nature is that of John Muir, the Scottish-American naturalist and environmentalist. Muir's deep love for the wilderness and his efforts to preserve natural landscapes led to the establishment of national parks in the United States. His writings and activism inspired countless individuals to appreciate and protect the natural world, highlighting the profound impact that one person's connection with nature can have on society.

The concept of biophilia, coined by biologist E.O. Wilson, suggests that humans have an innate affinity for nature and living organisms. This chapter will delve into the scientific evidence supporting biophilia and the various benefits of spending time in nature, including reduced stress, improved mood, and enhanced creativity.

Nature-based therapies, such as forest bathing (shinrin-yoku) and horticultural therapy, have gained popularity in recent years for their ability to promote physical and mental well-being. We will explore these practices and the ways in which they harness the healing power of nature to foster connection and holistic health.

The role of nature in fostering connection is not limited to individuals; it also extends to communities. Community gardens, urban green spaces, and conservation efforts bring people together, creating opportunities for collaboration and social bonding. This chapter will highlight the importance of communal engagement with nature in building strong, resilient communities.

4

Chapter 4: Creative Expression as a Means of Connection

Creative expression is a powerful tool for fostering connection, both with oneself and others. Engaging in artistic activities such as painting, writing, music, and dance allows individuals to communicate their emotions, thoughts, and experiences in unique and meaningful ways. This chapter will explore the various forms of creative expression and their impact on connection and well-being.

One inspiring story of creative connection is that of Frida Kahlo, the Mexican painter known for her vivid and emotive self-portraits. Kahlo's art was deeply intertwined with her personal experiences, and her ability to convey her inner world resonated with audiences around the globe. Her work serves as a testament to the power of creative expression in forging connections and fostering empathy.

The process of engaging in creative activities can also facilitate connection by providing opportunities for collaboration and shared experiences. Whether it's participating in a community theater production, joining a choir, or attending a writing workshop, creative endeavors bring people together and create a sense of belonging and camaraderie.

Art therapy is a therapeutic approach that harnesses the power of creative expression to promote healing and connection. By encouraging individuals to explore their emotions and experiences through art, art therapy provides a safe and supportive environment for self-discovery and growth. This chapter will delve into the principles and benefits of art therapy and its role in fostering connection.

The digital age has also transformed the landscape of creative expression, offering new platforms and opportunities for connection. Social media, online communities, and digital art forms have made it easier than ever for individuals to share their creative work and connect with like-minded individuals. We will examine the impact of digital technology on creative expression and the ways in which it has expanded the possibilities for connection.

5

Chapter 5: Connection Through Shared Experiences

Shared experiences are a cornerstone of human connection, providing opportunities for bonding, empathy, and mutual understanding. Whether it's celebrating milestones, overcoming challenges, or simply enjoying everyday moments together, shared experiences create lasting memories and strengthen relationships. This chapter will explore the importance of shared experiences in fostering connection.

One poignant story of connection through shared experiences is that of the Lost Boys of Sudan, a group of young refugees who fled their war-torn homeland in search of safety. Their arduous journey and the bonds they formed along the way highlight the resilience of the human spirit and the power of connection in the face of adversity.

Travel and adventure also offer unique opportunities for connection, as they expose individuals to new cultures, perspectives, and experiences. From backpacking through foreign lands to embarking on a road trip with friends, travel fosters a sense of curiosity, empathy, and shared discovery. This chapter will explore the transformative power of travel and its role in building meaningful connections.

Rituals and traditions play a significant role in fostering connection, as they provide a sense of continuity, belonging, and identity. From holiday celebrations to cultural ceremonies, rituals create opportunities for individuals to come together, reflect on shared values, and strengthen their bonds. We will examine the importance of rituals in cultivating connection and preserving cultural heritage.

In the modern world, virtual experiences and online communities have become increasingly important sources of connection. Online gaming, virtual reality, and social media offer new ways for individuals to connect and share experiences, transcending geographical and cultural boundaries. This chapter will explore the potential of virtual experiences to foster connection and the importance of balancing online and offline interactions.

6

Chapter 6: The Role of Empathy in Connection

Empathy is a fundamental component of human connection, enabling individuals to understand and share the feelings of others. The ability to empathize fosters compassion, trust, and mutual support, creating a foundation for meaningful relationships. This chapter will delve into the science of empathy and its role in fostering connection.

One inspiring story of empathy comes from the life of Princess Diana, who used her platform to advocate for marginalized and vulnerable populations. Her compassionate and empathetic approach to humanitarian work resonated with people around the world and left a lasting legacy of kindness and connection.

The practice of active listening is a crucial aspect of empathy, as it involves fully engaging with and understanding another person's perspective. This chapter will explore strategies for improving active listening skills and the ways in which they can enhance empathy and connection in our relationships.

Empathy is not limited to human interactions; it can also extend to our relationship with animals and the environment. Stories of animal rescues, conservation efforts, and environmental activism highlight the powerful impact of empathy in creating a more compassionate and connected world. This chapter will explore these stories and the ways in which empathy fosters a sense of responsibility and stewardship for the planet.

Developing empathy requires conscious effort and practice. Mindfulness, perspective-taking, and compassionate communication are valuable tools for cultivating empathy and enhancing our connections with others. This chapter will provide practical strategies for building empathy and creating more meaningful and supportive relationships.

7

Chapter 7: Connection and Personal Growth

Personal growth and self-improvement are deeply intertwined with our ability to connect with others. Through our relationships and interactions, we gain valuable insights into ourselves and the world around us. This chapter will explore the ways in which connection fosters personal growth and the importance of nurturing healthy relationships for self-development.

One inspiring story of personal growth through connection is that of Malala Yousafzai, the Pakistani activist for girls' education. Despite facing immense adversity, Malala's unwavering commitment to her cause and her ability to connect with people around the world have made her a powerful symbol of resilience and empowerment. Her journey illustrates the transformative power of connection in fostering personal and social change.

Mentorship and supportive relationships play a crucial role in personal growth. Whether it's a teacher, coach, or friend, having a mentor who believes in us and encourages our potential can make a significant difference in our lives. This chapter will highlight the importance of mentorship and the impact of positive relationships on personal development.

Connection also involves embracing vulnerability and authenticity. By being open and honest with ourselves and others, we create opportunities for deeper and more meaningful connections. This chapter will delve into the importance of vulnerability in building trust and fostering genuine relationships.

The process of personal growth is a lifelong journey, and connection serves as a guiding force along the way. This chapter will provide practical tips for nurturing connections and fostering personal development, emphasizing the importance of self-awareness, empathy, and resilience.

8

Chapter 8: Overcoming Barriers to Connection

Despite the inherent human desire for connection, various barriers can hinder our ability to form and maintain meaningful relationships. These barriers can include social anxiety, cultural differences, and past traumas. This chapter will explore the common obstacles to connection and provide strategies for overcoming them.

One remarkable story of overcoming barriers to connection is that of Temple Grandin, an autistic professor of animal science and advocate for individuals with autism. Grandin's unique perspective and determination to connect with both animals and humans have made her a trailblazer in her field. Her story demonstrates the power of perseverance and the ability to overcome challenges to forge meaningful connections.

Communication is a key component of connection, and misunderstandings can often arise due to differences in communication styles. This chapter will explore the importance of effective communication and provide tips for navigating diverse communication preferences to build stronger relationships.

Cultural differences can also pose challenges to connection, as varying customs, beliefs, and values can create misunderstandings and conflicts. This chapter will highlight the importance of cultural competence and provide strategies for fostering cross-cultural connections and understanding.

Healing from past traumas and negative experiences is essential for building healthy connections. This chapter will explore the impact of trauma on relationships and provide guidance on seeking support and practicing self-care to overcome these barriers and create meaningful connections.

9

Chapter 9: Connection and Community Building

Connection extends beyond individual relationships; it also plays a vital role in building strong and resilient communities. This chapter will explore the importance of community connection and the various ways in which individuals can contribute to creating a sense of belonging and support within their communities.

One inspiring story of community connection is that of the Harlem Children's Zone, a non-profit organization dedicated to improving the lives of children and families in Harlem, New York. Through comprehensive support services and community engagement, the organization has created a network of care and opportunity for its members. This story highlights the transformative power of community connection in fostering positive change.

Volunteering and civic engagement are powerful ways to build community connections and contribute to the greater good. This chapter will explore the benefits of volunteering and provide practical tips for getting involved in community service and activism.

Building inclusive and equitable communities requires intentional effort and a commitment to social justice. This chapter will delve into the importance of inclusivity and provide strategies for fostering diverse and equitable community connections.

The role of technology in community building is evolving, with online platforms and social networks providing new opportunities for connection and collaboration. This chapter will examine the potential of digital tools to enhance community engagement and the importance of balancing online and offline interactions.

10

Chapter 10: Connection and Well-Being

The link between connection and well-being is well-established, with strong social bonds contributing to mental, emotional, and physical health. This chapter will explore the various ways in which connection promotes well-being and the importance of nurturing relationships for overall health.

One compelling story of connection and well-being is that of the Blue Zones, regions of the world where people live significantly longer and healthier lives. Research has shown that strong social connections and a sense of community are key factors contributing to the longevity and well-being of individuals in these areas. This chapter will delve into the lessons we can learn from Blue Zones and the importance of social connection for a healthy life.

The practice of mindfulness and meditation can enhance our ability to connect with ourselves and others, promoting greater well-being and emotional resilience. This chapter will explore the benefits of mindfulness and provide practical tips for incorporating mindfulness practices into daily life to foster connection and well-being.

Physical activity and movement also play a role in fostering connection and well-being. Whether it's participating in team sports, joining a dance class, or simply going for a walk with a friend, engaging in physical activity can strengthen social bonds and improve overall health. This chapter will highlight the importance of physical activity in building connections and promoting well-being.

The role of gratitude in fostering connection and well-being cannot be overlooked. Practicing gratitude helps us appreciate the positive aspects of our relationships and enhances our overall sense of happiness and fulfillment. This chapter will explore the benefits of gratitude and provide practical strategies for cultivating a grateful mindset.

11

Chapter 11: Connection in Times of Crisis

During times of crisis, the need for connection becomes even more pronounced. Whether it's a natural disaster, a global pandemic, or personal hardships, the support and solidarity of others can provide comfort and strength. This chapter will explore the importance of connection in times of crisis and the ways in which individuals and communities can come together to support one another.

One powerful story of connection in times of crisis is that of Fred Rogers, the beloved television host known for his show "Mister Rogers' Neighborhood." Throughout his career, Rogers emphasized the importance of kindness, empathy, and connection, particularly during challenging times. His legacy serves as a reminder of the power of connection to provide hope and healing.

The role of mutual aid networks in times of crisis has gained attention in recent years, as communities come together to provide support and resources to those in need. This chapter will explore the principles of mutual aid and the ways in which individuals can contribute to building resilient and supportive communities during times of crisis.

The impact of crisis on mental health and well-being cannot be underestimated. This chapter will provide strategies for coping with stress, anxiety, and grief during challenging times, emphasizing the importance of seeking support and maintaining connections with loved ones.

Acts of kindness and generosity can have a profound impact on individuals and communities during times of crisis. This chapter will highlight inspiring stories of compassion and solidarity, demonstrating the powerful role that connection plays in fostering resilience and hope.

12

Chapter 12: The Future of Connection

As we look to the future, the ways in which we connect with one another will continue to evolve. This chapter will explore the emerging trends and technologies that are shaping the landscape of human connection and the potential opportunities and challenges they present.

One exciting development in the realm of connection is the rise of virtual and augmented reality, which offers new possibilities for immersive and interactive experiences. This chapter will examine the potential of these technologies to enhance connection and create new forms of social interaction.

The importance of digital literacy and responsible technology use will be crucial as we navigate the future of connection. This chapter will provide practical tips for using technology mindfully and fostering healthy digital habits to maintain meaningful connections.

The role of connection in addressing global challenges, such as climate change, social inequality, and public health, will be increasingly important. This chapter will explore the ways in which individuals and communities can come together to tackle these issues and create a more connected and sustainable world.

Ultimately, the future of connection will be shaped by our collective efforts to build a more compassionate, inclusive, and empathetic society. This chapter will emphasize the importance of nurturing our connections with ourselves, others, and the world around us, as we work together to create a brighter future.

13

Chapter 13: Connection Through Storytelling

Storytelling is a powerful means of connection that transcends time and culture. From ancient oral traditions to modern-day literature, stories have the ability to convey emotions, share experiences, and build bridges between individuals and communities. This chapter will explore the role of storytelling in fostering connection and the ways in which narratives shape our understanding of the world.

One captivating story of connection through storytelling is that of Maya Angelou, the acclaimed poet and author. Angelou's autobiographical works, such as "I Know Why the Caged Bird Sings," resonated with readers worldwide, highlighting the transformative power of personal narratives. Her storytelling not only connected her with audiences but also inspired others to share their own stories.

Storytelling is a universal language that transcends cultural and linguistic barriers. Folktales, myths, and legends from around the world offer valuable insights into the human experience and foster a sense of shared heritage. This chapter will delve into the significance of these traditional narratives and their role in building cross-cultural connections.

The digital age has revolutionized storytelling, providing new platforms for individuals to share their narratives. From blogs and podcasts to social media and online forums, digital storytelling offers unprecedented opportunities for connection and community-building. This chapter will examine the impact of digital storytelling and the ways in which it has expanded the possibilities for connection.

Personal storytelling, whether through memoirs, journaling, or spoken word, allows individuals to reflect on their experiences and connect with others on a deeper level. This chapter will highlight the therapeutic benefits of personal storytelling and provide practical tips for harnessing the power of narrative to foster connection and self-discovery.

14

Chapter 14: Connection and Creativity in the Workplace

reativity and connection are essential components of a thriving workplace. Collaborative environments that foster open communication, innovation, and teamwork not only enhance productivity but also contribute to employee well-being and job satisfaction. This chapter will explore the importance of connection and creativity in the workplace and provide strategies for cultivating a positive and innovative work culture.

One inspiring story of workplace connection and creativity is that of Pixar Animation Studios. Known for its groundbreaking films and collaborative culture, Pixar has created a work environment that encourages creativity and fosters strong connections among its employees. This chapter will delve into the principles and practices that have made Pixar a model for innovation and teamwork.

Creative problem-solving and brainstorming sessions are valuable tools for fostering connection and generating new ideas. This chapter will provide practical tips for facilitating effective brainstorming sessions and creating a collaborative atmosphere that encourages diverse perspectives and creative thinking.

The role of leadership in fostering a connected and creative workplace cannot be overstated. Effective leaders inspire trust, promote open communication, and encourage innovation. This chapter will explore the qualities of effective leaders and provide guidance on cultivating leadership skills that enhance connection and creativity.

The importance of work-life balance and employee well-being in fostering a connected and creative workplace will also be emphasized. This chapter will highlight the benefits of flexible work arrangements, wellness programs, and supportive policies that prioritize employee well-being and create a positive and connected work environment.

15

Chapter 15: The Legacy of Connection

The impact of connection extends beyond the present moment, leaving a lasting legacy for future generations. The relationships we build, the communities we nurture, and the actions we take to foster connection contribute to a more compassionate and interconnected world. This chapter will explore the enduring legacy of connection and the ways in which individuals can create a positive and lasting impact.

One inspiring story of a lasting legacy of connection is that of Nelson Mandela, the South African anti-apartheid revolutionary and political leader. Mandela's unwavering commitment to justice, reconciliation, and unity not only transformed South Africa but also left a profound impact on the world. His legacy serves as a powerful reminder of the enduring power of connection and the importance of striving for a more just and connected society.

Acts of kindness and generosity have the potential to create a ripple effect, inspiring others to pay it forward and contribute to a more compassionate world. This chapter will highlight stories of individuals and communities who have made a lasting impact through their acts of kindness and the ways in which these actions have fostered connection and positive change.

The role of education in fostering connection and creating a lasting legacy will also be explored. Educators who inspire curiosity, empathy, and collaboration in their students contribute to the development of future generations who value and prioritize connection. This chapter will highlight the importance of education in shaping a more connected and compassionate world.

Ultimately, the legacy of connection is built through the choices we make and the actions we take to nurture our relationships and communities. This chapter will provide practical tips for creating a lasting legacy of connection, emphasizing the importance of empathy, compassion, and intentionality in our daily lives.

The Art of Connection: Psychology, Nature, and the Creative Spirit

The Art of Connection delves deep into the essence of human bonds, exploring how psychology, nature, and creativity intertwine to shape our relationships and sense of belonging. Across fifteen captivating chapters, this book invites readers on a journey through the myriad ways we connect with others, the world around us, and ourselves.

The book begins by examining the fundamental importance of connection, highlighting the psychological and biological underpinnings that make social bonds essential for our well-being. It then ventures into the rich terrain of the natural world, uncovering how our relationship with nature can rejuvenate our spirits and strengthen our connections.

From the inspiring story of Jane Goodall's work with chimpanzees to the transformative power of Viktor Frankl's experiences during the Holocaust, *The Art of Connection* is filled with compelling narratives that illustrate the universal language of empathy and compassion. Readers will also discover the role of creative expression in forging connections, as exemplified by the life and art of Frida Kahlo.

In addition to exploring individual connections, the book delves into the dynamics of community building, the impact of technology on our relationships, and the importance of rituals and traditions in fostering a sense of belonging. Chapters on empathy, personal growth, and overcoming barriers to connection provide practical insights and strategies for nurturing meaningful relationships.

The journey continues with a look at how connection promotes well-being, especially during times of crisis. The legacy of connection is examined through the lens of influential figures like Nelson Mandela, whose commitment to justice and unity left an indelible mark on the world.

Finally, *The Art of Connection* envisions the future of human bonds, exploring emerging technologies and global challenges that will shape the way we connect in the years to come. This book is a celebration of the profound and enduring power of connection, offering readers a roadmap to building a more compassionate, inclusive, and empathetic world.